Made In Maine

AN HISTORICAL OVERVIEW

by Paul E. Rivard, Director

THE MAINE STATE MUSEUM

Publication Design by
Donald Bassett, Museum Graphics Designer

Printed by Gannett Graphics on
Maine-Made Paper

Photography by
Gregory Hart, Museum Photographer

THE GREAT SEAL OF THE STATE OF MAINE

—INTRODUCTION

Writings about the history of Maine have long been dominated by stories of the forests and the sea. These two principal aspects of the State's geography and natural history have impressed themselves deeply upon the popular perception of Maine: a land of lobstermen and lumberjacks, of salt water and tall trees. Poets, philosophers, historians, and casual visitors have all tended to view Maine through similar bifocals. In the far view they have seen the Maine that is woods. This is the Maine of the Allagash, the "Haynesville Woods," the "Airline" from Bangor to Calais, the Maine of lumbermen, hunters, campers, fishermen, guides, lumber trucks and forest rangers. In the nearer view they have found the other Maine, that which is seacoast. Here is a world of seamen and sailing ships, the Maine that is today a tourist mecca, awash each summer in clam chowder, out-of-State cars, and lobster bibs.

These are the two enduring popular views of Maine—the far and the near, the forests and the sea. There have been, by contrast, few views of Maine from the middle distance.

The woods and the ocean continue to dominate public appreciation of Maine today in the same way that these characteristics have nearly overwhelmed the telling of Maine's history. This is for good reason. The forests, which first provided incentive for settlement on the "eastern frontier," did indeed prove a source of both work and wealth throughout Maine's history. Similarly, Maine's long coastline, indented with harbors and connected to large inland water systems, promoted an economy of commercial trade. The harvesting of raw materials for shipment often surpassed the instincts toward agriculture and manufacturing. Unquestionably the woods and the ocean each contributed mightily to the "Maine difference."

Resorting to simple statistics, the historian can plainly enough demonstrate the unique historical development of Maine. While all of the states in northern New England enjoyed a similar historical development, Maine was distinguished in a remarkable way by the influences of the forests and the sea. In 1850, for example, there were some 4500 lumberjacks employed in the Maine woods, a number ten times greater than could be found in either New Hampshire or Vermont. Likewise, Maine households sent over 13,000 men to sea at mid-century, exceeding the combined total of its two northern neighbors by a factor of over 2000%. There can be little question about it—the trees and the ships have made Maine different.

This view of Maine, seen as a place of timberlands and ocean-going commerce, was aptly summarized in eloquent lines written by Henry David Thoreau:

> There stands the city of Bangor, fifteen miles up the Penobscot, at the head of navigation for vessels of the larger class, the principal lumber depot on this continent with a population of twelve thousand, like a star on the edge of night, hewing at the forest of which it is built...and sending its vessels to Spain, to England, and to the West Indies for her groceries....

Images of trees and ships have deservedly colored the view of Maine's past and they continue to define and characterize the State as a unique place to live and work. By contrast, it has remained a great deal more difficult to appreciate the complex, if commonplace, history of the rest of Maine's people—the history of the vast majority of Mainers who were not lumberjacks, not lighthouse keepers, not the captains of tall ships. Behind the Maine of romance and

Broom makers at the Portland Broom Company pose around 1870.

legend—the Maine of poets, artists, and "downeast" writers—there has always existed another Maine, one composed of an unseen population of farmers, laborers, fish packers, manufacturers, mechanics, artisans, and factory workers of all sorts. Here is another Maine history, one played out not in the woods or at sea, but rather in countless homes, workshops, mills and factories which have existed throughout Maine communities. This is a story that might lack the unique qualities for which Maine is famous. It is also a history somewhat short on romance. Nevertheless, this is an important story of many Maine people.

If much of Maine's history is laced with romance and myth, this is perfectly alright. Myth-making is an important way in which any society defines its values and aspirations. The story of the woods and the coast defines Maine for many people, and also provides a special identity for the State as a whole. But historians (and history museums) must assuredly distinguish between romance and reality, revealing the one without discarding the other. Failing in this, too much of the State's history will go unappreciated, the lives of too many Mainers forgotten. The interpretive purpose of the "Made in Maine" exhibition program is to broaden the parameters of popular State history to encompass a broader cross-section of Maine's people.

Evidence of a more varied and complex past can be found in virtually every town in the State—even in the most unlikely vacation spots along the coast. The community of Camden, for instance, is today one of the premier summer resort areas of central Maine, and for good reason. It is a handsome coastal town blessed with a beautiful harbor and imposing hills. Visitors to this town a little over a century ago would have found much that is the same as today—the harbor full of sailing vessels, the seagulls overhead, the moon over the water. But, during a short walk along Megunticook stream (which empties into Penobscot Bay at Camden) they would have found a great deal more than this; the walk would have revealed much about another Maine story. Immediately above the picturesque falls at harborside, the visitor would have found Carleton's grist mill followed by Moody's cracker factory where dough was kneaded by machines using water-power. A few more yards upstream was the mill of the Bryant marble works which specialized in polishing tombstones, and the "Collins Glass Cylinder Pump" factory which made glass cylinder pumps (whatever they were). Near the cracker factory there was also the Farrar carriage works and the extensive factory of Johnson, Fuller & Company which specialized in the manufacture of felt used in paper-making and was, in fact, the largest such factory in the entire United States. Spare parts for the machinery used in many of these establishments could be made by the nearby machine shop of Knowlton & Company. Camden in 1870 had the sea, the little stores, and the clam chowder, but it had much more. Its workshops, mills and factories speak of another Maine past, one that is easily overlooked today.

The Camden example is a reminder that, for most of Maine's 19th-century population, life was hard-earned in a variety of workplaces, larger and smaller, where there were produced a very wide range of commodities from sawn lumber to stoves; from toothpicks to locomotive engines; from polished tombstones to "glass cylinder pumps." As the 19th century progressed, such products of Maine manufacturies provided employment for tens of thousands of families—people whose history has somehow been lost along

This is a print of the engine "Forest State" made by the Portland Company around 1852. The Portland Company produced nearly 900 locomotive engines from the early 1850s to the 1890s.

the bifocal line that divides Maine into the traditional, but simplistic, sea/forest dichotomy. An important story waits still be to be told—a story not only of large industrial cities such as Biddeford, Lewiston, Millinocket, Rumford, Sanford and Portland, but also many thousands of small workshops, mills, and even homes scattered throughout the State. This Maine story encompasses shipyards and sawmills, which were prominent factors of the State's economy and landscape. It is, however, a story expanded to include cotton factories, shoe "shops," iron furnaces, and countless other industries.

Above is a rare interior view of Samuel Lancaster's chair "factory" in New Sharon, circa 1880.

This four-horsepower stationary steam engine was built by the J.W. Penney Company of Mechanic Falls in 1882.

If the purpose of the Maine State Museum is to help engender and sustain a respect and appreciation for Maine's past, then there is an obligation to accord equal respect and concern for the seamstress as for the shipwright, for the loom fixer as for the lumberjack. This Maine story is not only one of sawyers and ship carpenters; it is also one of jack spinners, doffers, puddlers, pattern makers, machinists, seamstresses, bonnet makers, card tenders, weavers, millers, founders and shoemakers.

A fine painting, depicting the launching of a Maine-built ship in 1810, easily suggests itself as a symbol of Maine craftsmanship, useful in illustrating the cover of a book such as this one. It is much harder to illustrate (or even acknowledge) that in 1810 the most important of Maine's manufactures, in terms of the finished product value, was not shipbuilding (which ranked second) but rather the weaving of cotton and woolen cloth in Maine homes. Revealing these largely hidden aspects of Maine history is a difficult task. This is one challenge addressed in "Made in Maine."

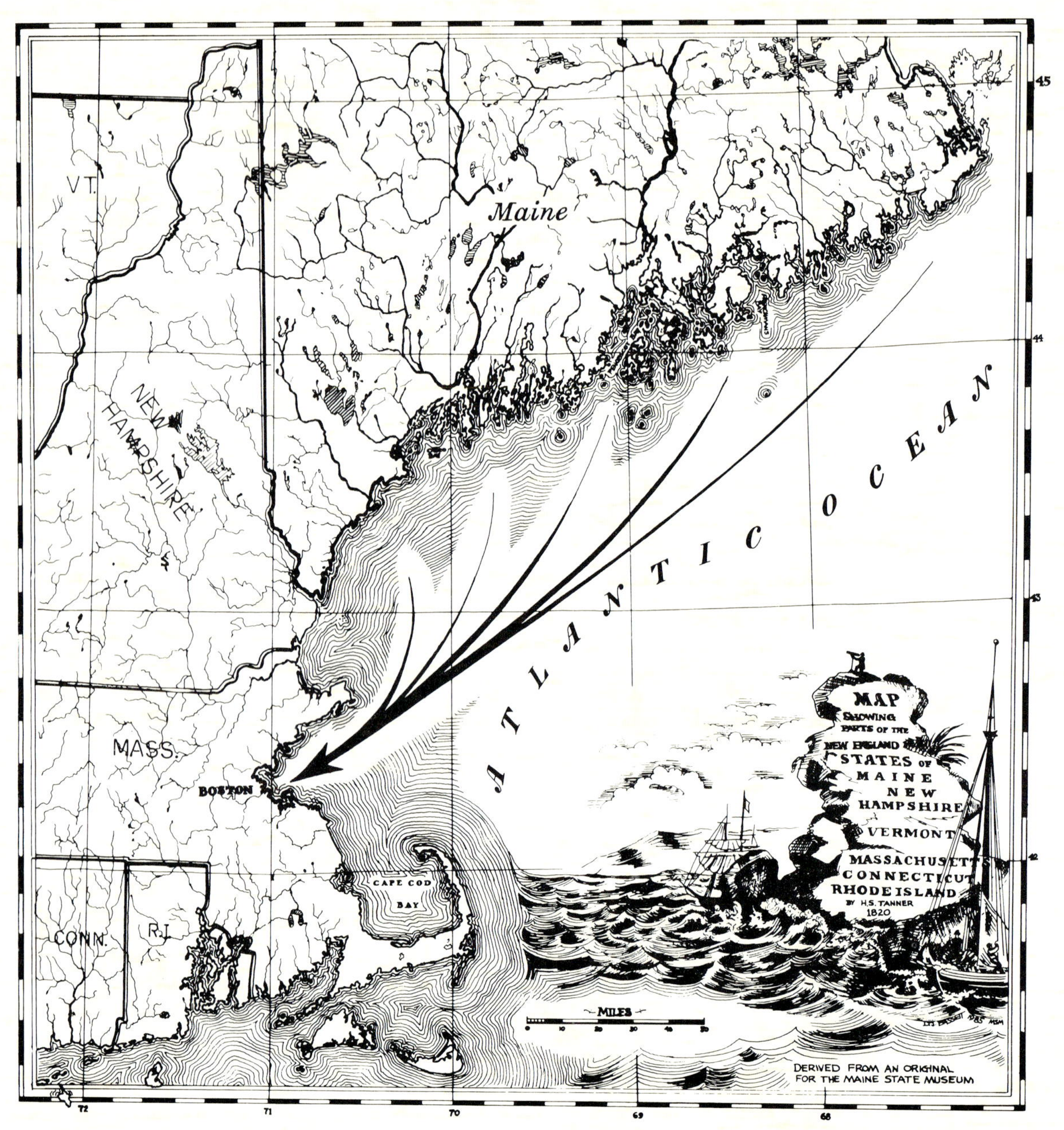

Maine
VT.
NEW HAMPSHIRE
MASS.
BOSTON
CONN.
R.I.
CAPE COD BAY
ATLANTIC OCEAN
MAP SHOWING PARTS OF THE NEW ENGLAND STATES OF MAINE NEW HAMPSHIRE VERMONT MASSACHUSETTS CONNECTICUT RHODE ISLAND
BY H.S. TANNER 1820
MILES
DERIVED FROM AN ORIGINAL FOR THE MAINE STATE MUSEUM
45
44
43
42
72
71
70
69
68

THE MAINE STORY

A glance at a map of New England reveals the relatively short distance that lies between Maine's coastline and the region's principal port of Boston. While Maine settlements were initially carved from a rugged frontier, few early Maine communities were really isolated; few were truly remote from the centers of trade. Coastal and trans-Atlantic shipping brought to Maine not only "her groceries," but also a substantial supply of manufactured commodities. As a consequence, Maine families were seldom far from a supply of essential goods, or even luxuries if they could afford them. The accessibility of most Maine towns to coastal trade diminished the incentive to manufacture goods. Many Mainers simply found that harvesting the State's abundant raw materials—timber, lime, granite—was a more convenient activity, and one which required little capital investment and faced few risks from competition.

In the 19th century, however, manufacturing activity in Maine was encouraged by several of the State's environmental characteristics. The first of these was the abundance of natural harbors and the network of navigatable rivers and estuaries which facilitated trade. Ships that carried products into Maine, and raw materials out, could also be used to supply manufacturers and to carry Maine-made products to distant markets. A second advantage was the large number of water power sites available in Maine. The State's inland water system encompassed literally thousands of likely mill and factory sites which could be developed at relatively little expense. The rapid development of power machinery during the 19th century made Maine's water power resources increasingly attractive to manufacturers and invited investment in Maine industries, especially as many choice sites in Massachusetts and Rhode Island became overdeveloped. Finally, Maine offered a reservoir of industrious men and women, scattered throughout the State's households, who were anxious to earn steady wages.

The value of this labor pool must not be underestimated as an incentive for the establishment of larger manufacturing plants in Maine. When the State itself sought to promote additional industrial investment in the 1860s, officials argued clearly, and directly, that "workers of the best character and in any required number can be procured from amongst the sons and daughters of our farming population; accustomed to labor, robust, trustworthy and intelligent, and glad for employment within their native State...." This, to be sure, was a statement of pure salesmanship, but it was also largely true. In the 19th century, as today, a compelling justification for industrial development in Maine was the work ethic of its people.

Trade opportunities, water power, and hard-working people were factors which each contributed to the great increase of manufacturing which took place in Maine during the 19th century. The large factories built along some of Maine's major rivers in Biddeford and Lewiston stand today as obvious reminders of this manufacturing growth. But beyond these, there was a countryside peppered with smaller mills and workshops, with literally thousands of smaller workplaces where Maine men and women labored to produce a great variety of manufactured goods both for use in Maine and for sale to the nation and the world.

Among the small workplaces so important to the story of Maine were many that have left virtually no mark upon recorded history. This is especially true of the home environment. While homes were obviously built to serve primarily as family residences, they could easily double as workplaces in which portable tools—such as spinning wheels, hand looms, and sewing machines—were set to work attending to family needs or making goods for sale.

During the 18th century and the first decades of the 19th, the principal manufacturing activities in Maine homes were the spinning of yarn and the weaving of cloth. More than 16,000 Maine homes provided space for hand looms in 1810, and an even greater number had spinning wheels in use. Sometimes the textiles produced with these tools were used within the family itself, but they were also sold or traded locally as an item of commerce. Since the annual production of cloth in Maine homes exceeded both family and local needs, this manufacture was clearly beyond the scope of routine housekeeping and became a significant item of trade and a source of employment in homes. More affluent households often employed others to spin and weave on a contract basis. Many more-or-less professional spinners and weavers earned their board by doing this work in neighborhood homes, or even working on contract for distant yarn spinning mills.

Entrepreneurs and merchants were quick to realize the potential of the skilled labor force available in Maine homes. The system of "putting out" raw materials for manufacture or assembly in households was extended during the 19th century to include not only cloth, but bonnets, and the sewing of clothing. Through their work in these "cottage" industries, Maine family members contributed substantially to household income and, simultaneously, to the State's overall economy. Textile factories ultimately brought an end to the era of hand weaving in the home, but, in return, they provided an increasingly large supply of less expensive cloth. This promoted an expansion of family wardrobes, increased attention to style and fashion, and provided more work (often in the home) for tailors and seamstresses.

This view of a wool spinner was taken at Center Lovell in 1902. Although woolen factories had provided machine-spun woolen yarn for generations, the use of spinning wheels often persisted as a domestic tradition. Wool used in the above spinning demonstration was carded at the nearby "Hapgood" carding mill in Waterford.

The sewing of clothes rapidly took the place of hand spinning and weaving as a home manufacturing activity. Initially the clothing was sewn laboriously by hand, but the invention and perfection of sewing machines increased productivity tremendously and helped to create an even greater market for clothing. From the mid-19th century onwards, nearly every community employed at least one professional tailor or seamstress, and a great many communities had several. The primary articles produced were custom women's clothing sewn to outfit the local community with the fashions of the day. In the manufacture of men's clothing, however, Maine households were destined to play an even greater part. Beginning in the 1840s, the sewing of many items of men's wear was centralized through the development of "ready-to-wear" garments requiring little custom tailoring. (Mid-19th century photographs of Maine men, even prominent and wealthy men, illustrate a certain lack of concern for tailoring.) The trade in these garments was controlled by Boston merchants who designed and pre-cut the parts for countless thousands of shirts, pants, vests and coats. To sew these parts into finished garments, the manufacturers looked to Maine homes. As a direct result, tens of thousands of Maine women were employed sewing garments in their own households between the 1860s and the 1880s. The number of garments made in Maine ran to the millions. By 1880 the number of Maine women employed in this way had reached such a high level that the U.S. Census was forced to appoint a special commissioner just to report on this extensive household employment.

Through the making of cloth, and the sewing of clothing, household manufacturing touched many thousands of Maine homes. But these domestic trades are often overlooked because, in part, the work was integrated on a daily basis with household chores and the making of things for strictly family use. The work of Maine's spinners, weavers, and seamstresses have provided few visible memorials. Today the looms and sewing machines have been moved out of the living rooms of Maine homes and the textiles and garments have, themselves, been mostly worn out and discarded.

The table-mounted knitting machine shown here was manufactured by the Hinkley Knitting Machine Company of Bath in the 1860s. The development of machines like this followed closely the perfection of sewing machines, which were also made in Maine.

Gardiner was a leading site for the manufacture of stoneware pottery during the 19th century. Shown above is a covered jar made by Ballard & Brothers, 1855-1856.

The manufacturers of artisan workshops are not so easily forgotten. Shop environments produced most of the cherished "antiques" which are avidly collected as romantic souvenirs of the past. Pewter teapots, tin boxes, Windsor chairs, silver spoons, brass andirons, flintlock pistols, cane fishing rods, painted mirrors, tall clocks, shoes, barrels and bureaus etc.—all of them were products of the artisan workshop. Here craftsmen produced a large body of yesterday's necessities and today's "collectibles" which fill antique shops and history museums.

When Maine became a State in 1820 there were already some 1700 artisan workshops in the former "Province" of Maine, in addition to countless repair shops located on family farms. As the decades of the 19th century advanced, the number and variety of workshop manufactures continued to grow. Some, like blacksmith shops, were geographically dispersed throughout the State in direct service to local farming populations. In addition to the shoeing of horses and oxen, the work of these small country blacksmith shops encompassed the making and repair of farm tools, and often the forging of axles for local carriage builders. Because blacksmith shops provided custom work for a more-or-less local community, there needed to be a great many of them—there is a limit to the distance you can make an ox walk for a new shoe. In this regard, however, blacksmith shops were not particularly typical of artisan workshops in general.

The Franklin-type stove was made by Eleazer Wyer and Joseph Noble in Portland, 1822-1830.

During the 19th century the custom manufacture of goods declined steadily under the influence of mass-production techniques. Most of Maine's workshops produced a quantity of goods that far exceeded the needs of the strictly local community. The making of shoes provides an excellent example. In the 18th century Maine families customarily ordered their shoes custom-made by a local shoemaker. This practice became rapidly obsolete in the 19th century, however. Lasts, or "wooden feet," were established in regular sizes and substituted for real feet. This released the shoemaker from a dependence on the needs of the immediate community. The mass-production of ready-made shoes, produced to standard sizes, allowed the shoemaker to become a manufacturing entrepreneur. While women's shoes, and some more expensive men's shoes, continued to be sewn on custom order, the great expansion of the ready-made men's boots and shoes fostered a considerable growth in shoemaking in the number and productivity of Maine shoemaking shops. By the 1840s Maine shoemakers were producing many tens of thousands of pairs of shoes annually for sale in Boston, New York, the American south, and even the West Indies.

Because the products of so many artisan workshops were destined for sale in distant places, the best shop locations tended to be near easy water transportation or major interior highways. The community of Stevens Plains (now Westbrook) saw an uncommon concentration of artisans whose wares were broadly distributed inland aboard peddler's carts which operated on circuits leading as far as Quebec and the Canadian maritime provinces.

This ambrotype of a young shoemaker was made by Bangor photographer Merritt Jordan in the late 1850s or early 1860s.

The tinware containers shown at the left were made at Stevens Plains (now Westbrook) in the 1840s. This community was a regional leader in the manufacture and distribution of tinware which was often sold from peddlers' carts traveling long routes to northern Maine, Nova Scotia and Quebec.

During the first half of the 19th century, the workshops of Stevens Plains produced large quantities of small articles such as tinware, brass, pewter, brushes, combs, brooms, and pails. The manufacture of tinware alone illustrates in a compelling manner the level of productivity that could be achieved in relatively small workshops. In 1850 the tinsmiths of Stevens Plains produced no less than 107,000 tin pots, pans, pails, and other assorted tinware pieces. This manufacture (which employed only 25 men) was able to serve, even flood, an enormous region with tinware goods. This sort of workshop productivity, geared to a wider marketplace and organized to use mass-production techniques, was typical of many workshops in Maine, workshops that had little to do with local community needs.

Commerce allowed for an easy distribution of both raw materials and finished products, and to appreciate the manufactures of Maine's workshops it is necessary to see them in the larger context of regional trade.

Many Maine workshops used local materials, but many others required materials secured from other States, or even from abroad. It may come as a shock to collectors of "early American" ironwork that the enormous bulk of iron used by New England blacksmiths was imported from Russia and Sweden. Similarly, Maine stoneware pottery was made from New Jersey clay, and brooms made in Skowhegan used broom-corn raised in Illinois. Maine's abundant routes of commerce, its coastline, and its inland water traffic, opened the State's commercial doors to the entire world. Through this door passed every sort of commodity, raw material, or product part. Alas, the romantic vision of the "colonial craftsman," toiling at his bench making custom products for barter with local families, is not a very big part of the story of manufacturing in Maine.

Above left: A view of the Katahdin Iron Works, around 1875. Katahdin was the only furnace in Maine which produced iron from raw ore.

Below left: Lumber is stacked near an unidentified Maine sawmill (probably in the Skowhegan area).

Above right: The workshops of L.D. Hobbs, carriage ironworker; and D.M. Moody, carriage maker, were housed in the same building in Caribou, 1882-1883.

This exterior view of a water-powered mill was taken in Bar Harbor around 1870.

Maine craftsmen of the 18th century could, and did, produce a very wide range of consumer products using only simple hand tools and traditional skills and experience. But this situation changed dramatically in the 19th century as new, faster and smarter, machines were invented to perform many production tasks. The 19th century was an era distinguished by the advance of machinery of all kinds, and, for the most part, this machinery was driven by water power—Maine's cheapest and most plentiful energy source. The mechanization of many craft processes that took place during the century greatly diminished the importance of skill and training in the craftsman's work. In return, however, the machines promoted the building of new workshop industries along Maine rivers and streams. The age of machines helped to create new jobs in Maine.

It is impossible to guess accurately the number of water-powered mills that were built along Maine rivers during the 19th century, but the total certainly runs into the thousands. A survey of Maine's water power, undertaken by the State Government in the 1860s, counted some 5,100 rivers and streams on the map of Maine. These, the report noted, were fed by countless other smaller branches that "thread the State with a fine network of brooks and rivulets, so that in all parts it seems alive and in motion with running water—a distinctive and characteristic feature, which strikes the attention of the stranger and the traveller." Along these many rivers and streams, State officials identified more than 3,000 water power sites that were suitable for the building of mills and factories.

It is not particularly surprising to discover that Maine's water power resources were tapped first, and most frequently, by sawmills. In between the lumberjacks of the Maine woods on the one side, and the seamen and their ships on the other, ran the rivers of Maine descending, inexorably from the State's many lakes to the sea. On these stood the sawmills, employing thousands of sawyers, often leading settlement on the frontier, and providing a cornerstone of the State's economy. The earliest sawmills were built during the first decades of European settlement in Maine during the 1630s and 1640s; and by 1840 there were some 1400 sawmills in the State, each doing its part to transform the timberlands into marketable beams and boards, to be shipped around the world. In addition to millions of feet of "long lumber," new types of saws also made possible the production of clapboards, barrel staves, and shingles.

Maine's manufacture of these materials was impressive: in 1850 alone, for instance, the State's mills produced some 100,000,000 pine and cedar shingles.

The construction of sawmills often preceded or was totally independent of settlement, responding to commercial opportunities rather than neighborhood services. Other types of mills, serving the immediate needs of the local population, were built somewhat later—even grudgingly it sometimes would seem. Nevertheless, as the farming population grew, so did the investment in gristmills to grind locally-grown corn into meal, carding mills to prepare wool fibers for home spinning, and fulling mills to felt and treat woolen cloth woven in homes. These mills were built in support of local communities and, as such, they functioned much like blacksmith shops. By 1820 there had been built in Maine some 210 carding and 149 fulling mills. These statistics further demonstrate the importance of spinning and weaving in the State's households.

The State's kilns and furnaces also processed Maine raw materials into new raw materials and building parts. These included only brickyards, but iron furnaces and foundries, potteries and, particularly, lime kilns. The products of these industries ranged from barrels of lime destined for use in plaster, mortar, and cement, to cast iron stoves and machinery. The kilns of the Rockland area made lime burning the most important industry in mid-coastal Maine during the middle of the 19th century. The shipment of lime from Maine dominated the Boston and New York markets. Meanwhile, the most numerous of the furnace industries—iron foundries—became more closely associated with machine shops. Through this combination, they often grew into engineering plants. The largest of these, the Portland Company, produced a large variety of machinery and steam engines of all sorts, including nearly 1,000 locomotive engines built between 1852 and 1892.

While abundant water power opportunities fostered a proliferation of many small mills and factories scattered throughout the State, they also permitted the development of large factory complexes and, as a result, the creation of some of the State's largest cities. The harnessing of larger water powers, like those along the Saco and Androscoggin rivers, involved an extensive engineering and construction of dams, gates, sluiceways, and turbine installations. Since this entailed an expense not warranted in the building of sawmills, gristmills, and smaller woolen factories, these sites had been largely avoided during the first part of the 19th century. The development of these large water power sites awaited the investment of bigger manufacturers, and the first industries to make these investments were those making cotton and woolen cloth. Their investments in Maine began in the 1820s and gained momentum in the 1850s and 1860s as the manufacturers of southern New England looked northward in search of power and labor. The wholesale development of large water powers, and the building of entire factory towns, was a new phenomenon for Maine, one which was first seen along the Saco river. Here, in a region which had once been the State's chief outlet for timber and sawn lumber, the Saco Manufacturing Company demonstrated a capacity for factory construction not previously seen in the State. In the mid 1820s, this company began the construction of an ambitious wooden factory building which, for a brief few months, was thought to be the largest textile factory building in the world. This wooden factory burned down in 1829 and its replacement was built (like the house of the most frugal of the three pigs) of bricks. The construction of factories in the State led to a new and greatly expanded need for bricks; and in the Saco/Biddeford area alone there arose some dozen brickyards by 1850.

These cotton factory buildings were built along the Androscoggin River at Lewiston Falls beginning in the early 1850s. By 1870, Lewiston boasted some 15 factory complexes like this one.

The emergence of Lewiston as the State's premier textile manufacturing center came nearly a quarter-century after the development of the first Saco factories. At mid-century there was only one small woolen factory in Lewiston, and the population was only around 4,000. A concerted, and throughly planned, water power development was launched in 1850, however, and this quickly led to the incorporation of a number of cotton and woolen factories. The Bates factory began weaving cotton cloth in 1852, and it was joined, the following year, by the two factories of Lewiston Mills. The Hill company began in 1854, the Franklin in 1857, the Androscoggin in 1860, and the Continental in 1866. By the late 1860s, some 15 distinct factories had been built along the Androscoggin below Lewiston Falls and these employed over 3,500 women and 1,600 men. The population of the city had grown five-fold to more than 21,000 and the construction of the factories had necessarily been matched by the construction of dwellings to house this great influx of workers.

The impact of the textile industry in Maine is most easily seen, even today, in the Saco-Biddeford area and in Lewiston. However, the stories of these cities are only the most noticeable because they were concentrated in relatively small areas. Elsewhere in the State numerous other cotton and woolen factories were built and these worked to shift local economies away from the traditional standbys of lumbering, shipbuilding, and agriculture and toward the manufacturing economy familiar to the rest of New England. Along with the obvious cotton and woolen factories, the State saw a great increase in other

This rare interior photograph shows master weaver Kenneth Daggett in the "weave shed" of the North Monmouth Woolen Mill around 1914.

This interior view of a Maine shoe "shop" is believed to have been taken in Portland around 1870.

industries (such as foundries and machine shops) which produced an increasing supply of machinery ranging from steam engines to knitting machines. Furthermore, it was through the influence of ever-increasing machine technology that other manufactures were drawn to the water power and the labor pool of Maine. This was particularly true of shoemaking.

The mass-production of shoes had long been centered in eastern Massachusetts where it was controlled by merchants and guilds of skilled workmen. New machinery, however, permitted the relocation of manufactures to any place where labor could be found and trained. Many shoe manufacturers looked toward northern New England, and particularly to Maine. The relocation of this industry to the north was remarkable; local Maine communities sensed the benefits of manufactures to the local economy, and vied openly for the location of "shoe shops" in their communities, offering such incentives as free land, free water power, and exemption from local taxes for a period of time. Towns even offered to construct the needed buildings; shoemaking magnates from Massachusetts could, in fact, come shopping in Maine for the most desirable location for their enterprises. They did. As a result, Maine emerged dramatically in the 1860s to 1880s as a center of shoemaking, and a number of communities became "shoe towns," the largest of which, with shoe production reaching 4,000,000 pairs per year, was Auburn.

The workers shown here were employed by the Ayer-Houston hat factory in Portland in the 1870s.

A significant change in the State's economy took place during the second half of the 19th century as Maine participated in regional, even national opportunities provided by mechanization and industrialization. In spite of the continued importance of sea and forest products, Governor Selden Connor was led to recognize in 1877 that the value of manufactured products had recently doubled that of the State's agricultural output, and that the value of cotton goods alone had eclipsed that of sawmill products. While shipbuilding remained an important industry in the State, the Governor reported that its products were now surpassed not only by the value of goods made in cotton and woolen factories, but also by the output of ironworks, tanneries, boot and shoe factories, and even flour mills.

The dominance of textiles, iron and shoes in the Maine economy was relatively short-lived, however. Although some woolen factories utilized wool raised in Maine, and shoe making promoted the State's tanneries, a great many of the larger industries of the State were not truly indigenous to the State and did not utilize raw materials found in Maine. Most of these enterprises were financed by out-of-State investors and were located in Maine primarily for its water power and its labor. These manufactures have, in the 20th century, been free to leave the State in response to 20th century economic considerations, and many have. The machinery of the 19th century was, like its products, largely generic, and portable. In Maine, however, the forests and the sea always remained to provide a special enduring backbone for the economy and safety-valve for the workforce.

The forests and the sea, while temporarily eclipsed by the 19th century industry, each remained a seemingly never-failing resource. The forests, the enduring forests, which had earlier produced the lumber trade, similarly gave rise to the Maine paper industry once chemistry permitted the manufacture of paper from wood pulp. The forests which had built the lumber city of Bangor, also built the paper city of Millinocket, and they did so dramatically. While

Wall mural featuring documented Maine crafts people. By: Museum Graphics Designer, Donald J. Bassett.

Millinocket consisted of little more than a farmhouse and a pasture in 1898, just two years later it was producing some 240 tons of newsprint a day and had spawned a community of 2,000 people. Today, this product of the Maine woods is the State's most important manufacture. Meanwhile, the tradition of shipbuilding has grown, particularly at Kittery and Bath, to once again rank as the State's second-largest manufacturing industry. In the end, it was again the forests and the sea that had the most to say about Maine's unique development.

One irreversible by-product of 19th century industrial development in Maine was the urbanization caused by the insatiable demand for factory labor. The growth of manufacturing interests, which characterized the second half of the 19th century, quickly depleted Maine's available pool of labor—or, at least, absorbed those individuals who were prepared to work for factory wages. This shortage of labor led, in turn, to increased immigration into Maine. Labor-hungry industries promoted a resettlement of workers which has forever transformed, and enriched, the ethnic mixture of Maine. Although this is most noticeable today in the larger textile towns that were poplated by immigrants from French-speaking Quebec province, there are numerous other examples which might be cited, from the Irish and English ironworkers brought to work at the Pembroke Iron Works to a concentration of Armenian immigrants hired by the J.T. Winslow pottery company in Portland. In Maine, as elsewhere throughout the United States, an industrial economy brought with it new employment opportunities which, in turn, attracted new people. From this development there emerged a new Maine demography in which men and women with surnames such as Gilroy, Cavanaugh, Pelletier, Dionne, Everson, and many others took their places in "downeast" homes and workshops alongside the Jewetts, Blaisdells, Jordans, Prebles and Moodys.

As compared with the phenomenal development of the major rivers of Rhode Island and Massachusetts, the national impact of Maine's manufactures was not particularly great. But it was, nevertheless, significant, and from the viewpoint of the Maine people whose history we must tell, manufactures were often critical to their very existence in Maine at all.

Theirs is the story of Maine which has long been obscured behind the veil of romance which has curtained off these more mundane aspects of history and left us, for too long, with a vision of Maine as composed of forests and sea alone. This, then, is the story of another Maine; one composed of men and women sewing coats, casting stoves, sawing shingles, dressing spinning machines, fashioning fine silver, painting tinware, sewing bonnets, making oilcloth, weaving carpets, felting cloth, turning bobbins, shoeing oxen, carding wool, and making fishing rods. It is this story which the "Made in Maine" exhibition hopes to tell.

BUILDING THE EXHIBITION

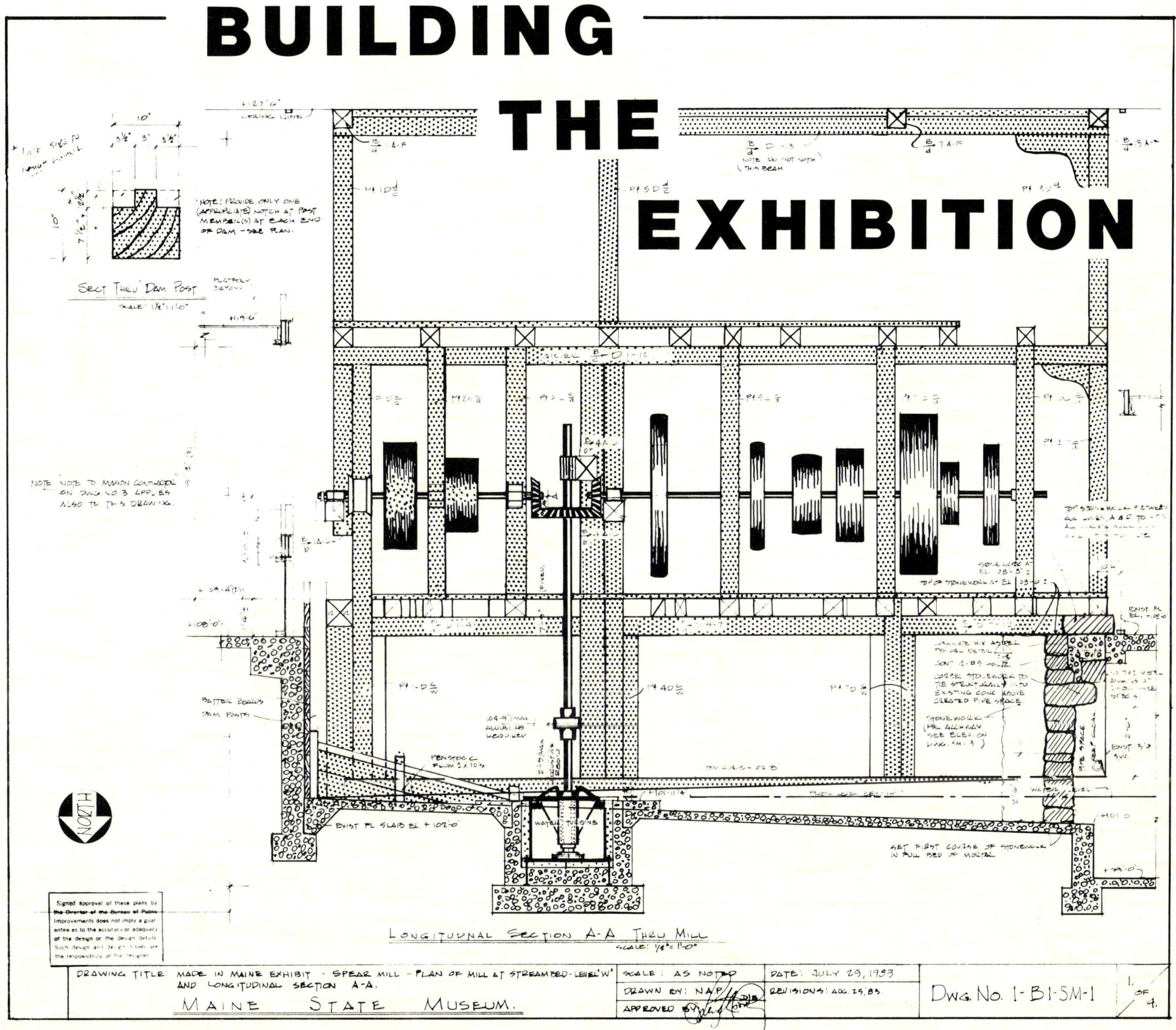

STATE · OF · MAINE
LIBRARY · MUSEUM · ARCHIVES

— BUILDING THE EXHIBITION

Since 1971, when the "Cultural Building" housing the Maine State Museum was first opened to the public, the Museum staff has planned and built a number of sophisticated exhibitions. These have included a series of diorama-style natural history installations, and numerous history exhibitions employing diverse artifacts ranging from a 30-ton piece of a ship's hull to a Revolutionary War potato peeling! The Museum has amply demonstrated its capacity to plan and build substantial installations, using artifacts that are both enduring and ephemeral. "Made in Maine" has been, however, the largest and most complex installation yet attempted by the Maine State Museum. It is the result of a sustained research, planning, collecting, and construction program which, itself, has been the most ambitious effort in the institution's history. Four years of continuous research and planning preceded the groundbreaking early in 1983. The exhibition, originally expected to require five years of construction, was completed nearly 2½ years ahead of schedule for an opening in the fall of 1985.

Exhibitions built by the Maine State Museum (or for that matter by any serious history museum) should reflect directly the important collections available for display. Collections are, after all, the central reason for the museum's existence. So, the first step leading to the construction of "Made in Maine" was an extensive review of the Museum's artifact holdings. This was done in 1979. The question asked in this analysis was simple: "What important artifacts does the Maine State Museum have in its collections?" This exercise reinforced the belief, and reminded the staff, that the Museum's most numerous and significant collections fell into three broad areas: (a) archeological specimens, (b) artifacts made in the State of Maine, and (c) tools used by Maine workers and Maine industry. These findings were not, certainly, a surprise; the Museum had, for more than a decade, directed its attention to these particular collecting areas. This analysis, nevertheless, provided guideposts for future exhibit development at the Museum. In particular, a plan was developed to build two large long-term installations on the lower level of the Cultural Building. These were termed "Made in Maine" (dealing with the products of Maine manufacture and related tools and machines), and "12,000 Years in Maine" (encompassing the early cultural history of Maine from the Pleistocene "ice age" through the first stages of European exploration and settlement). Due to the size of many of the most improtant artifacts involved, it was apparent that "Made in Maine" would require the greatest amount of space and entail the biggest construction "mess" yet seen inside the Museum building. The staff decided to build this exhibition first.

A basic exhibit "theme" was initially sketched out and this was followed by a research project which identified the most important aspects of Maine's manufacturing history. This analysis also recommended those portions of the story best suited to the installation space. The list of options considered was much longer than was first anticipated, and it became clear that the Museum could never hope to reflect more than the basic outlines of the overall story, using representative "samples" to express the greater whole. First consideration was given to the importance of certain industries to Maine. At the same time, however, the staff identified artifacts available for exhibit use. It was very quickly shown that much of Maine State Museum's artifact collections were not ideally suited to a superior installation of this particular theme. While the initial exhibition concept paralleled the Museum's collections of Maine-made things, it was apparent that a great many additional historical items would be needed. The Museum, therefore, embarked upon an unprecedented effort, between 1979 and 1983, to locate and secure both typical and rare tools, machines, and manufactured products that were "made in Maine." This effort fostered a dramatic growth of the Museum's collections, and influenced the planning of the exhibition in a very direct way. Many exhibit ideas were reinforced by the Museum's discovery of rare and important artifacts suited to telling the "Maine" story. This was true of the woolen factory exhibits, for instance—installation which became possible only after the acquisition of some very early and rare pieces of wool processing equipment.

This view of a carriage maker's shop is unidentified but is believed to show one of several similar carriage manufactories in Union around 1880.

The second story of the Museum's recreated woolen factory features machinery used for carding wool fibers, spinning yarn, and winding yarn for later use in the weaving process. The yarn spinning "jack," shown in the foreground of the above view, was made by Gilbert, Gleason & Davis of North Andover, Massachusetts . . . the supplier of similar machinery to most Maine woolen mills from the 1830s onward.

Shown above and to the right are views of the "Spear" woodworking shop as it appeared in Warren. The center section of this building was built in the 1760s by Moses Copeland, and later enlarged and modernized by Isaac Spear and his descendants beginning around 1850.

As the building had deteriorated to the point of near collapse, the Museum staff carefully dismantled the entire structure, numbered the several parts, and moved them to the Museum in Augusta for reconstruction as the centerpiece of the exhibition. The exhibition features the shafting and pulleys (shown at right), together with the water turbine running in a flowing stream of water.

Some installation ideas were necessarily abandoned when no significant artifacts could be found. This was true of the paper-making industry. The Museum found that the smallest of papermaking machines were, in fact, larger than the entire Cultural Building itself! Another problem faced the proposed shoe factory installation. There were simply no "attics" full of early shoemaking machinery that could be found.

Other likely installations were avoided, in the planning process, because they were prominently shown in other Maine museums, or even in other installations at the Maine State Museum itself. This was true of both sawmilling and shipbuilding, which already were integral parts of an earlier installation dealing with Maine's extractive industries.

Ultimately, an exhibit plan was crafted to adapt to the stories that were important in terms of Maine history, and also permit exhibition of some of the Museum's best objects. In addition to the woolen machinery mentioned above (looms for weaving, spinning and carding machines, and "fulling" stocks used to felt cloth), the Museum discovered and acquired a large number of important artifacts to support other "Made in Maine" installation ideas. Among these was a three-story woodworking shop found in Warren, Maine. This building, complete with its 19th century water turbine, line shafting, and gearing, was dismantled piece by piece and moved to the Museum for later installation in the exhibition. Meanwhile, machinery for use in this shop was secured from yet another woodworking shop in Woodstock, Maine.

Shown above and below are views of Museum staff members Norman Payne and Paul Bonin. Careful documentation of each part led to the preparation of a detailed layout (shown at the beginning of this section) drawn by Norman Payne.

The Whiting "No. 6" cupola casting furnace used by the Portland Stove Foundry Company was moved to the Museum early in the construction work. It would now be impossible to move an artifact this large into the Museum without cutting a new opening in the outside wall.

The larger artifacts and building fragments helped suggest design options while influencing the thematic organization of the installation. Acquisition of some large artifacts, like the "cupola" furnace used by the Portland Stove Foundry Company, was only initiated when it was obvious that space would be available to make adequate use of it.

As the planning proceeded, the collection of artifacts grew dramatically in both quantity and quality. Many items acquired were products of Maine's numerous workshop manufactures: pewterware, tinware, pottery, furniture, silver, and other related decorative arts. Tools and machinery needed to express the story of Maine's working population were more difficult to find; factories seldom have allocated valuable manufacturing space to the dead storage of old, obsolete, machinery. The Museum's success in securing these artifacts ultimately surpassed even the most optimistic of early hopes.

Exhibits Preparator Robert Barnard puts some finishing touches on the casting furnace.

The staff realized that the exhibition could only actually show a small percentage of the different manufactures of the State. Efforts were directed toward devising an organizational structure in which sample installations might be used to express a larger group of similar manufactures. A woolen factory, it was suggested, might reasonably represent a range of similar factory operations including cotton factories, shoe factories, and indeed all other manufacturing activities organized on a factory basis. Efforts to arrange the State's manufacturing enterprises in a simple way led ultimately to the organization of the exhibit theme into four separate groups of "working" environments: the "home," the "shop," the "mill" (including the furnace industries), and the "factory." This approach was over-simplified, to be sure. It proved to be, nevertheless, a useful mechanism for imposing order on a complex story.

The idea of recreating "environmental" settings was an attractive one from the start. By using complete or partial architectural reconstructions, the role of the curatorial staff in exhibit planning was greatly increased, and this was considered desireable. Furthermore, the preparation of full-scale room scenes was one familiar to museums and held the promise of including many more artifacts than would customarily be utilized in a more formal exhibit design. The proposed organization of the installation according to environmental groups was critical to both the unfolding of the exhibit theme and to the planning and construction process which followed. The exhibit design which resulted permitted the use of over 2000 artifacts from the Museum's collections.

Most of the work environments recreated by the Museum were faithfully copies from actual Maine sites. The sewing room installation was based on a similar room in an East Madison home. From this home, the Museum also acquired the sewing machine, sewing patterns, and other related artifacts.

The Museum's plan of organization for the installation was reinforced by a series of logos drawn by Gerald Blouin of South China.

While there were a number of home-based, or "cottage" industries that were historically important in Maine, the most significant were the making of textiles and the sewing of clothing. An 1820s kitchen scene and an 1880s parlor scene were selected to represent the era of hand spinning and weaving, and the era of home sewing, respectively. There were, necessarily, a greater number of installations devoted to workshop manufactures, since a great variety of workshops were found in the State and the Museum had collections representing many of them. Installations were planned to show the workshops of a machinist, blacksmith, cabinetmaker, gunsmith, shoe maker, and fishing rod manufacturer. The Museum also installed the complete water-powered woodworking shop which had earlier been dismantled and moved from Warren. Two installations were chosen to represent the mill and furnace industries: an 1830s fulling and cloth finishing mill, and the foundry furnace from Portland. To represent the factory group, the Museum recreated a small two-story woolen factory with carding, spinning, and other machinery on the upper level, and a suite of seven power looms in the "weave shed" below. The basic design of "Made in Maine" was composed of these basic elements, plus a number of display cases and platforms for the exhibition of Maine-made products.

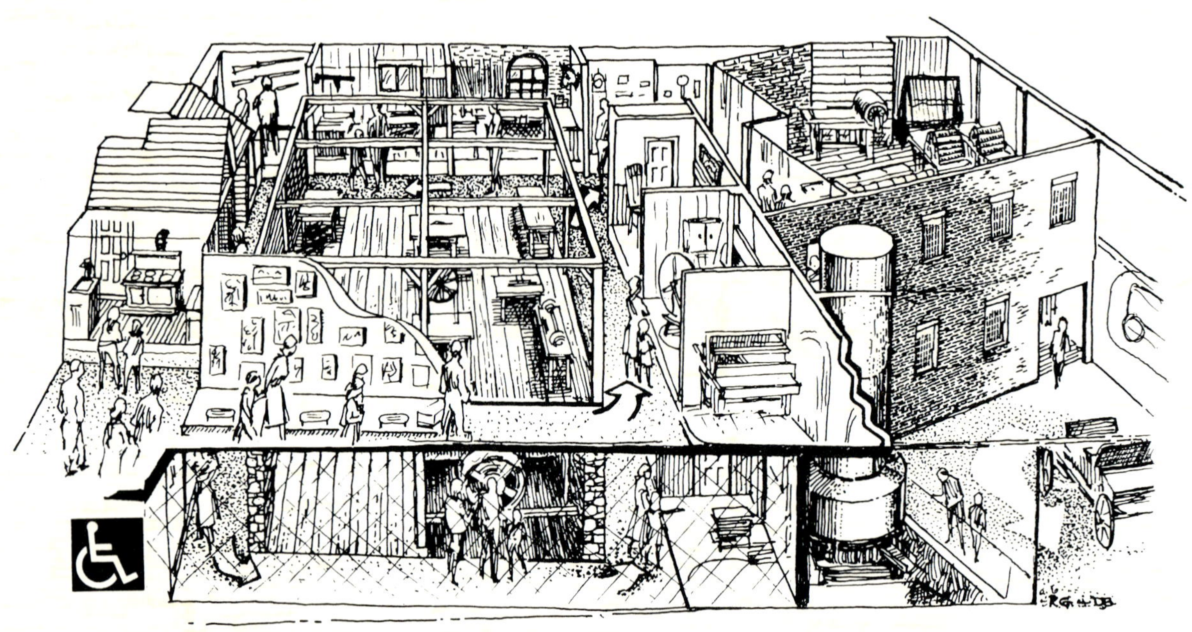

The cut-away drawing shown above was produced to show the multi-level complexity of the installation. Drawing by Roger Gould of Manchester.

Following an extensive examination of these exhibit ideas, and the successful search for needed artifacts, an aggressive design and engineering phase was possible. The challenge was to arrange the large number of disparate installations noted above into a relatively limited space inside the Museum building. A design plan was needed which would make sense interpretively, display the important collections, and accommodate the needs of the visiting public. The Museum's design staff provided a solution which utilized the entire volume of the space available, not just the floor area. The water-powered shop was placed in the center of the space and a rampway was designed to encircle the installation. This proved to be a superior solution which provided an unfolding perspective of the workings of the water-power system, allowed numerous spaces on the periphery for other installations, and permitted access for the handicapped throughout.

Decisions concerning the interior layout of the environmental scenes were made by the curatorial staff in concert with the design effort. Rather than seeking to create "typical" environments, one or more sites were selected to serve as direct examples to copy. The woolen factory, thereby, came to be patterned after a woolen factory built by the Samuel Mayall Company of Gray in 1834; the blacksmith/machine shop emerged as a direct copy of a small workshop located on the Collins family property in Manchester, etc. An engineering analysis of the entire installation was prepared—a study which led to the preparation of blueprints and contract documents needed for major new excavations, steelwork, framing, electrical, plumbing and mechanical systems.

Jack hammers and backhoes were brought into the Museum building to demolish most of the lower level floor in preparation for excavation, concrete, and steel work.

From the beginning, the staff dreamed that the turbine below the woodworking shop might actually be turning in a stream of running water to simulate actual operating conditions. The Museum embarked on a bold (but nerve-wracking) program to accomplish this, one which called for major alterations to the building itself. To install the complete shop structure, and provide the necessary stream bed, dam, sluiceway, and turbine in operation, it was necessary to enlarge the building—not by adding a wing, but rather by excavating within the building to a depth of some 16 feet below the existing floor level! From a construction perspective, the most striking aspect of the entire project (and certainly the noisiest and messiest) was the initial phase during which jack hammers, backhoes, trucks, and front-end loaders were at work inside the Museum, demolishing the existing concrete floor, excavating tons of gravel, sand and debris, and preparing for the concrete footings and the mill raceway.

Construction began early in 1983, and by fall the basic skeleton of cement and steel was in place. With progress on the installation proceeding remarkably well, the Museum advanced the completion date and, at the same time, made a decision to enlarge the exhibit area itself. The staff was reminded during the early phases of construction that the area immediately adjacent to the end of the exhibit provided an unusually high ceiling height. Here, then, was an opportunity to display some larger Maine-made products—artifacts which, for lack of space, the Museum had previously avoided collecting. The staff decided to seize the opportunity presented. As a consequence, the State's collections were further enlarged to include such objects as sleighs, carriages, boats, bicycles, automobiles, and engines.

The "Made in Maine" exhibition program, which had first been conceived to show the Museum's existing collections, ultimately provided motivation for a dramatic growth in the collections owned, and cared for, by the Maine State Museum. Many hundreds of important artifacts were added to the State's collection as a direct result of this project. In the long term, it may well be in its stimulation of collection acquisitions and care that "Made in Maine" will provide its most enduring benefits to the people of Maine.

Above: New steelwork was required to support the exhibit areas and to provide a long rampway descending around the "Spear" woodworking shop installation.

Below: Excavation and new concrete work was required throughout the installation.